AF270239

STEGOSAURUS

by Laura K. Murray

Consultant: Mathew J. Wedel, PhD
Western University of Health Sciences
Pomona, California

PEBBLE
a capstone imprint

Published by Pebble, an imprint of Capstone
1710 Roe Crest Drive, North Mankato, Minnesota 56003
capstonepub.com

Library of Congress Cataloging-in-Publication Data
Names: Murray, Laura K., 1989- author.
Title: Stegosaurus / by Laura K. Murray.
Description: North Mankato, Minnesota : Pebble, an imprint of Capstone, [2025] | Series: Dinosaur guides | Includes bibliographical references. | Audience: Ages 5-8 | Audience: Grades 2-3 | Summary: "Readers want to know all about dinosaurs! Dig into the facts on what made Stegosaurus different from other dinosaurs. Engaging text and images make this book a great choice for information to use in a report or just to read for fun"— Provided by publisher.
Identifiers: LCCN 2024020691 (print) | LCCN 2024020692 (ebook) | ISBN 9780756589165 (hardcover) | ISBN 9780756589240 (paperback) | ISBN 9780756589202 (ebook pdf) | ISBN 9780756589264 (kindle edition) | ISBN 9780756589257 (epub)
Subjects: LCSH: Stegosaurus—Juvenile literature.
Classification: LCC QE862.O65 M897 2025 (print) | LCC QE862.O65 (ebook) | DDC 567.915/3—dc23/eng/20240524
LC record available at https://lccn.loc.gov/2024020691
LC ebook record available at https://lccn.loc.gov/2024020692

Editorial Credits
Editor: Erika L. Shores; Designer: Dina Her; Media Researcher: Jo Miller; Production Specialist: Tori Abraham

Image Credits
Alamy: tony french, 27; Capstone: Jon Hughes, cover, 1, 5, 6, 12, 13, 14, 18, 22; Getty Images: Hulton Archive, 25 (right); Science Source: Mark Garlick, 11, RICHARD BIZLEY, 17; Shutterstock: Kues, background (throughout), Daniel Eskridge, 7, Elenarts, 21, Everett Collection, 25 (left), Stefano Carnevali, 24, tinkivinki, 8; Superstock: Francois Gohier/Mary Evans Picture Library, 28

Printed in the United States   6692

Table of Contents

Words in **bold** are in the glossary.

The Roof Lizard

What dinosaur had pointy plates on its back? Stegosaurus! Its name means "roof lizard." Scientists used to think the plates laid flat like shingles on a roof. But now scientists know they stuck up nearly straight.

Stegosaurus lived during the Late Jurassic Period. That was 152 to 145 million years ago.

Did You Know?
Stegosaurus disappeared about 80 million years before dinosaurs such as Tyrannosaurus rex lived.

Where in the World

Stegosaurus lived in the western United States and Europe. In the United States, it roamed Colorado, Utah, and Wyoming. In 2007, a Stegosaurus **fossil** was found in Portugal.

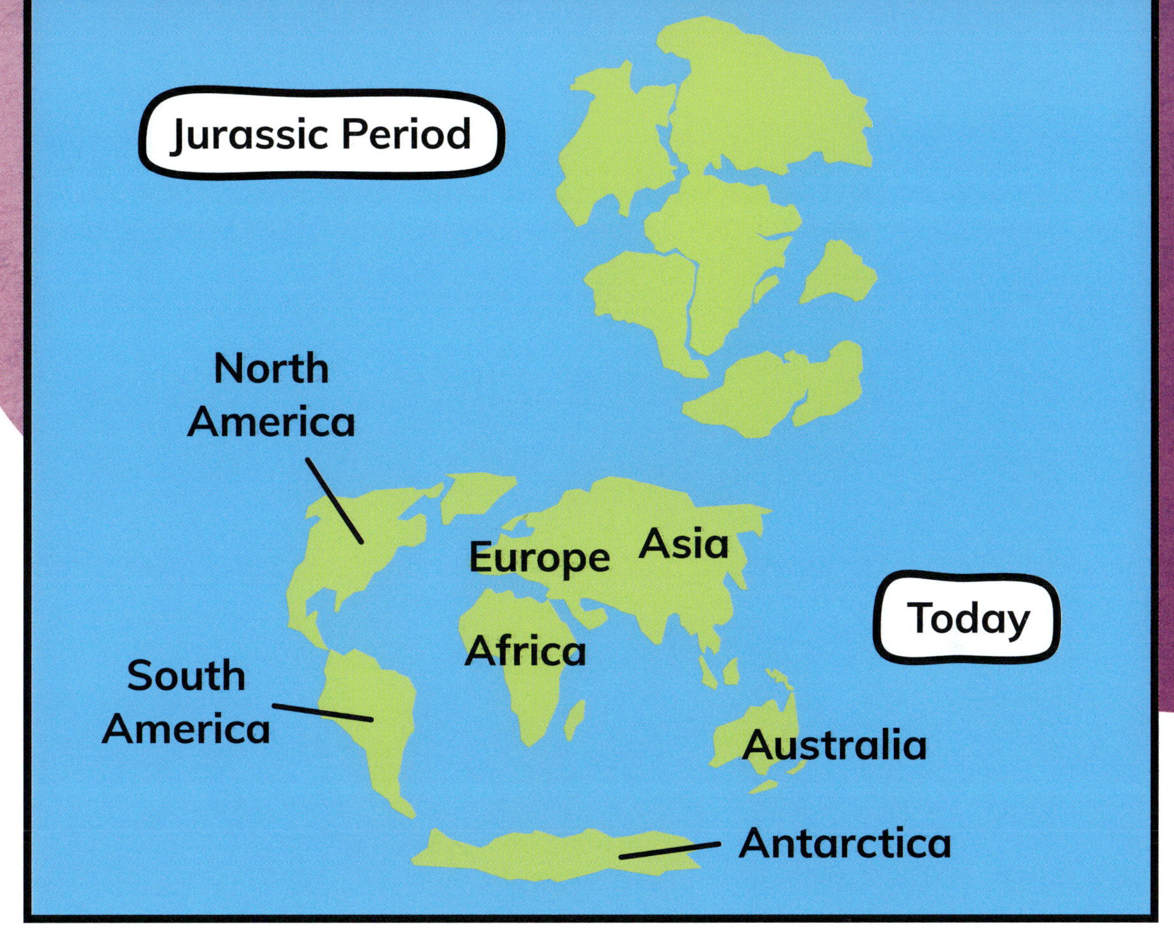

Earth's land was different when dinosaurs lived. The land began as one piece. Then it slowly broke apart. It took millions of years. The land was in two pieces at the start of the Jurassic Period. North America and Europe were joined. Then the land broke apart more.

Stegosaurus roamed flat, semi-dry
land. It needed a lot of plants to eat.
It likely lived in places by rivers or other
water. Plants would grow better there.

Earth did not have flowers or grass
at that time. The land had ferns and
seed plants. There were some trees too.

Stegosaurus Bodies

Stegosaurus had a large body. It grew up to 30 feet (9 meters) long. It moved slowly on four legs. Its back legs were longer than its front legs. Stegosaurus had a short neck with a small head.

Did You Know?

The brain of Stegosaurus was small. It was about the size of a lime.

Stegosaurus had a strong tail. It had four big, sharp spikes. The spikes were about 3 feet (0.9 m) long.

Stegosaurus likely used its tail during fights. It held its tail high in the air. It would swing the strong, spiked end at enemies.

plate

Stegosaurus had large plates sticking out on its back and tail. The plates were thin and flat. They were made of a bony material that contained **blood vessels**. There were 17 plates.

The plates may have been used for different things. They may have helped fight off **predators**. The plates may have helped the dinosaur stay warm or cool. They may have been used to show off to **mates**.

What Stegosaurus Ate

Stegosaurus ate plants. It did not eat other dinosaurs or animals. It likely ate plants and bushes that grew close to the ground. It could not reach far with its short neck.

Stegosaurus ate shrubs, moss, and ferns. It ate plants called cycads. Cycads grew millions of years ago, and they are still alive today. Cycads have cones, like pine trees.

Stegosaurus had about 78 small teeth. The teeth acted like scissors. They could cut and chop plants. Stegosaurus had a bite that was more powerful than other plant-eaters. It had a strong jaw for chewing, like a cow.

Did You Know?

Scientists compare Stegosaurus to living animals to study how it chewed its food.

Life of Stegosaurus

Stegosaurus may have lived in family groups. The dinosaurs may have traveled together in herds. But scientists do not know. They have not found enough information yet.

Scientists put Stegosaurus into three groups. The dinosaurs in the three groups are different sizes. Their spikes and plates are different shapes and sizes.

Stegosaurus could not move fast. Its top speed was 5 miles (8 kilometers) per hour. Its predators were larger dinosaurs. Allosaurus was a threat. This powerful dinosaur walked on two legs.

Scientists have found an Allosaurus bone with a hole in it. The hole likely came from a Stegosaurus tail spike.

Discovering Stegosaurus

Fossils help people learn about dinosaurs. The fossils may be things such as bones or footprints. They give clues about life millions of years ago.

In the mid-1870s, the first Stegosaurus bones were found in Colorado. In 1877, scientist Othniel Charles Marsh gave Stegosaurus its name. Marsh was in a race against scientist Edward Drinker Cope. They wanted to see who could discover more dinosaur bones. The race was called the "Bone Wars."

In 2003, fossil hunter Bob Simon made an important find in Wyoming. It was the skeleton of a young adult Stegosaurus. It was the most complete skeleton of the dinosaur ever found. It had 360 bones.

Scientists later named the skeleton Sophie. They used **lasers** and other tools. They made **3D** models of the bones. Scientists wanted to learn how the dinosaur ate, walked, and acted.

Sophie

Today, people study Stegosaurus. They explore and look for fossils. Sometimes they find plates and spikes! Scientists still have a lot of questions about how Stegosaurus lived. There is much more to learn about these amazing dinosaurs.

Fast Facts

Name: Stegosaurus (meaning "roof lizard")

Lived: Late Jurassic Period (about 152 to 145 million years ago)

Range: western United States (Colorado, Utah, Wyoming) and Portugal

Habitat: flat, semi-dry places

Food: plants such as ferns, moss, cycads

Threats: larger dinosaurs such as Allosaurus

Discovered: found in Colorado; named and described in 1877

Glossary

3D (THREE-DEE)—having width, height, and depth

blood vessels (BLUD VES-uhls)—a system that carries blood through the body

fossil (FA-suhl)—the remains or traces of a living thing from many years ago

laser (LAY-sur)—a tool that uses powerful light

mate (MAYT)—a partner that joins with another to produce young

predator (PRED-uh-tur)—an animal that hunts other animals for food

Read More

Finn, Peter. *The Mighty Stegosaurus*. New York: Enslow Publishing, 2022.

Throp, Claire. *Read All About Dinosaurs*. North Mankato, MN: Capstone, 2022.

Vonder Brink, Tracy. *The Stegosaurus*. New York: Crabtree Publishing, 2024.

Internet Sites

American Museum of Natural History: Stegosaurus
amnh.org/explore/ology/ology-cards/017-stegosaurus
-stenops

National Geographic Kids: Stegosaurus
kids.nationalgeographic.com/animals/prehistoric/facts
/stegosaurus

Natural History Museum: Stegosaurus
nhm.ac.uk/discover/dino-directory/stegosaurus.html

Index

About the Author

Laura K. Murray is the Minnesota-based author of more than 100 books for young readers. She loves learning from fellow readers and helping others find their reading superpowers! Visit LauraKMurray.com.